THE INCREDIBLE WORLD OF PLANTS

PLANTS OF THE DESERT

CHELSEA HOUSE PUBLISHERS
New York • Philadelphia

Text: Andreu Llamas
Illustrations: Luis Rizo

Plantas del desierto © Copyright EDICIONES ESTE, S. A., 1995. Barcelona, Spain.

Plants of the Desert copyright © 1996 by Chelsea House Publishers, a division of Main Line Book Co. All rights reserved.

1 3 5 7 9 8 6 4 2

Library of Congress Cataloging-in-Publication Data

Llamas, Andreu.
  [Plantas del desierto. English]
  Plants of the desert / [text, Andreu Llamas : illustra▮ Luis Rizo].
    p. cm. — (The Incredible world of plants)
  Includes index.
  Summary: Describes how desert plants adapt to their env▮ ment with its extreme temperatures, occasional down▮ almost constant drought, unrelenting sun, and shifting sa▮
  ISBN 0-7910-3466-6. — ISBN 0-7910-3472-0 (pbk.)
  1. Desert ecology—Juvenile literature. 2. Desert plants—▮ nile literature. [1. Desert ecology. 2. Desert plants. 3. Ecol▮ I. Rizo, Luis, ill. II. Title. III. Series: Llamas, Andreu. Incre▮ world of plants.
QH541.5.D4L5813   1996   95-1▮
581.5′2652—dc20

Contents

WHAT IS A DESERT?

Not all deserts are large expanses of hot, dry sand. There are also deserts of rock, salt, and even ice.

You probably think of the desert as a huge area of sand under the scorching sun. However, not all deserts are like that, since some are located in very cold regions of the planet. Not all deserts are made of sand, either—some are rocky, and others can be formed by salt or ice. But, all deserts do have one thing in common: they are regions that almost continuously suffer from drought, so there is very little water available. Some deserts have not received a single drop of water in 15 years!

The desert's surface is almost totally devoid of vegetation since the wind quickly carries it away.

Air in the desert has so little moisture that clouds do not form—that is why the desert sky is almost always clear and bright, which means that the ground is always exposed to the direct action of the sun's rays.

The air's temperature can reach 100 to 120 degrees Fahrenheit, and the surface temperature of the sand and rocks during the hottest part of the day can get up to 170 degrees Fahrenheit! The abrupt changes in temperature between day and night cause the rocks to break until they form fragments that are so small that the wind carries them away.

(1) Sand and stone
The desert climate is so rough that there is always an ongoing struggle for survival where life is almost impossible. Only 1/5 of the desert's expanse has vegetation, and the rest of the land is only sand and stone, or just rock.

(2) Cactus forest
Wild teasel sometimes form actual "cactus forests," where it would be impossible for people to walk, because their skin would be torn by all the thorns. However, these formations do house a large number of animals.

(3) Thorny nests
Everything is different in the desert, and all of the inhabitants must adapt. Where else could you find a dove's nest made of thorns on a cholla cactus?

(4) Globular forms
To fight the desert wind and lack of water, many plants adopt globular and tubular shapes and grow in a protected position, downwind from the dunes.

4

DESERT PLANTS

Deserts have more plants than you can imagine.

Life in the desert is difficult for plants, since they cannot run and hide in the shade or bury themselves underground like animals. The sun scorches, and temperatures change from infernal heat during the day to very cold nights. Also, rain is extremely scarce, and when it does come it is always violent and torrential.

In order to survive, desert plants have learned to be resistent, to wait, and to grow and reproduce quickly when rain does fall.

Despite these limitations, desert vegetation is low and sparse. In general, the plants have a dry, spiny look, and trees are very rare indeed (most of them belong to the acacia family).

To endure the driest season, desert plants use three clever tricks: (1) some plants lose their stems and leaves and survive solely through their roots, (2) other plants wither and leave their seeds waiting for the next rainfall, and (3) the true desert plants—the only ones that are permanently visible—have gone through great transformations in order to be able to conserve water (the cacti are the most well known of these).

In order to avoid having to share water with other plants, many species have developed toxic mechanisms to repel neighbors and fellow species and keep them off of their territory. The result is that the plants stay very far away from each other.

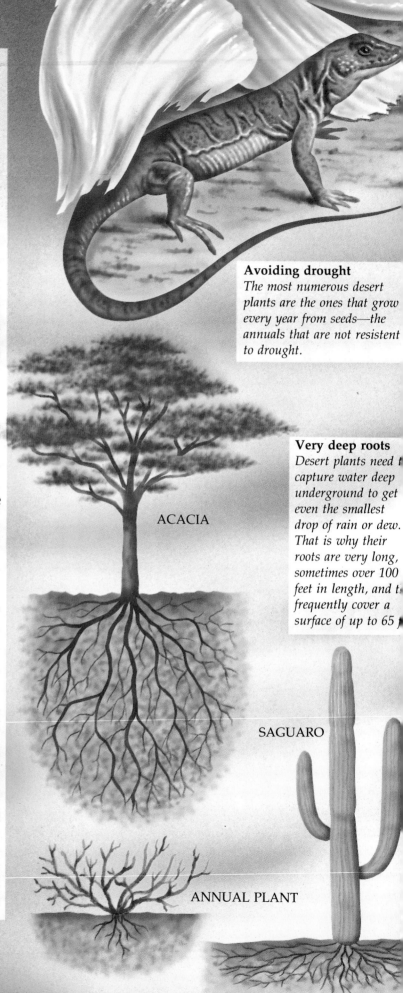

Avoiding drought
The most numerous desert plants are the ones that grow every year from seeds—the annuals that are not resistent to drought.

Very deep roots
Desert plants need t capture water deep underground to get even the smallest drop of rain or dew. That is why their roots are very long, sometimes over 100 feet in length, and t frequently cover a surface of up to 65 f

ACACIA

SAGUARO

ANNUAL PLANT

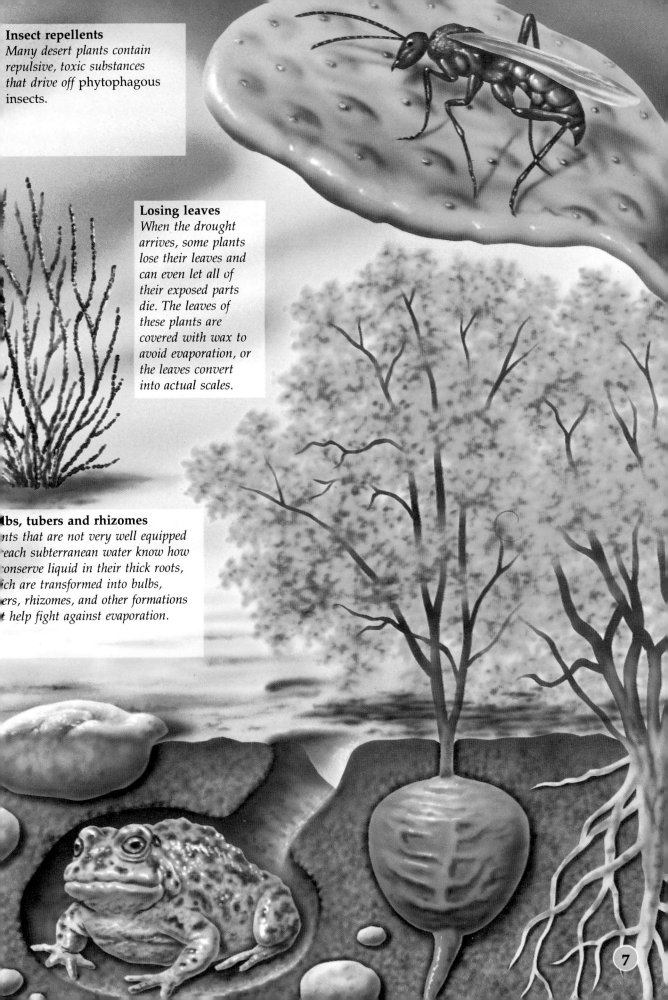

Insect repellents
Many desert plants contain repulsive, toxic substances that drive off phytophagous *insects.*

Losing leaves
When the drought arrives, some plants lose their leaves and can even let all of their exposed parts die. The leaves of these plants are covered with wax to avoid evaporation, or the leaves convert into actual scales.

lbs, tubers and rhizomes
nts that are not very well equipped reach subterranean water know how onserve liquid in their thick roots, ich are transformed into bulbs, ers, rhizomes, and other formations help fight against evaporation.

7

FIGHTING AGAINST THE SAND

The oceans of sand in the desert look like no other landscape on earth.

Because the sand is blown by the wind, the desert's surface changes constantly, like the sea, and the wind creates huge, strange waves, which are sand *dunes*. When an obstacle blocks the wind's force, the wind loses its power and drops the sand accumulated while it was blowing. This is how dunes are born, advancing because the sand constantly receives the wind's force.

How can plants survive in such a difficult environment? There are a special type of desert plants—perennial grasses—that reproduce thanks to the growth of new *shoots* in their system of roots. After a few years, these roots' networks become huge complexes, and the subterranean growth of the plant can be up to 85 percent of the whole plant. The great expanse of the roots permits the plant to capture water over quite an extensive area. During the driest months, the exposed vegetation dries out and dies, while underground, the subterranean root system stays alive.

These plants are very important in sandy deserts, since they help hold together the desert's surface, which is in constant motion. Many of these plants even start to grow roots from any part that is buried in the sand, so it is difficult for them to die entirely, since they only need some new roots to develop in order to begin to grow again.

(1) Impressive dunes
Dunes frequently form immense oceans of sand which, for example, take up 1/4 of the Sahara Desert. Dunes advance about 66 feet (20 meters) per year, and the biggest ones can be up to 980 feet (300 meters) high!

(2) Sand storms
Every year, the wind mobilizes 60 to 200 million tons of sand. During desert storms, the grains can be brutally lifted hundreds of feet into the air.

(3) Walking without sinking
The animals that live in the sand have developed different systems to walk the sand without sinking. This geckonid uses its fingers and toes to do this, while the horned rattlesnake (3a) advances side to side, moving its body like a lever.

(4) Roots everywher
As you can see, perennial grass fights stay above the desert's sand with its spectacular system of roots.

(5) Coral serpents
The very poisonous coral snakes dive into the sand searching for insects. They also inse themselves in the sand when it is too hot.

(6) Sand fish
They belong to the lizard family, but thei long, slippery, aerodynamic bodies le them slip easily throu the sand, swimming a if they were fish in water.

4

5

A THORNY LIFE

The authentic desert plants have had to adapt quite a lot in order to survive the scarcity of water.

One survival technique is to absorb all the water available and then store it in succulent tissue. Plants that store water like this are also called succulent, and the most well known of these belong to the *cactus* family, which includes more than 2,000 different species.

Cacti develop thorns instead of leaves. These thorns are very useful: first, they are hard and very sharp, to dissuade animals that might try to devour a cactus; secondly, they provide shade for the stem and collect the morning dew; and third, the thorns trap a layer of air around the plant to reduce the amount of evaporated moisture, forming a barrier against the hot, dehydrating air. Chlorophyll appears on the stems and on the exterior tissue of the plant, since there are no leaves. That way, *photosynthesis* can be carried out without wasting any water.

Cactus stems are bent into creases and crests, and they can fold and unfold like an accordion. Thanks to this ability they can store large quantities of water when it rains. Later, during the months of drought, the cactus shrinks more and more as it uses up the water stored inside.

New cacti spring up from the cactus bulbs that are underground, but they can also be formed from seeds, which can grow without being pollinated by a cactus flower of the same species.

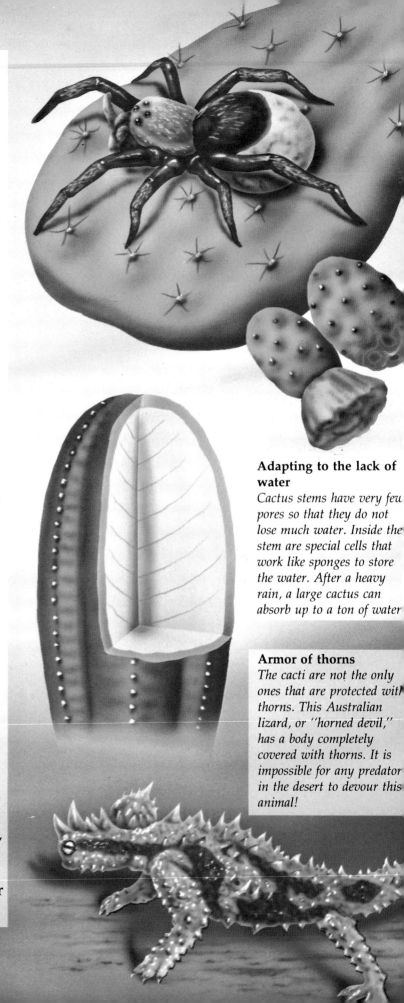

Adapting to the lack of water
Cactus stems have very few pores so that they do not lose much water. Inside the stem are special cells that work like sponges to store the water. After a heavy rain, a large cactus can absorb up to a ton of water

Armor of thorns
The cacti are not the only ones that are protected with thorns. This Australian lizard, or "horned devil," has a body completely covered with thorns. It is impossible for any predator in the desert to devour this animal!

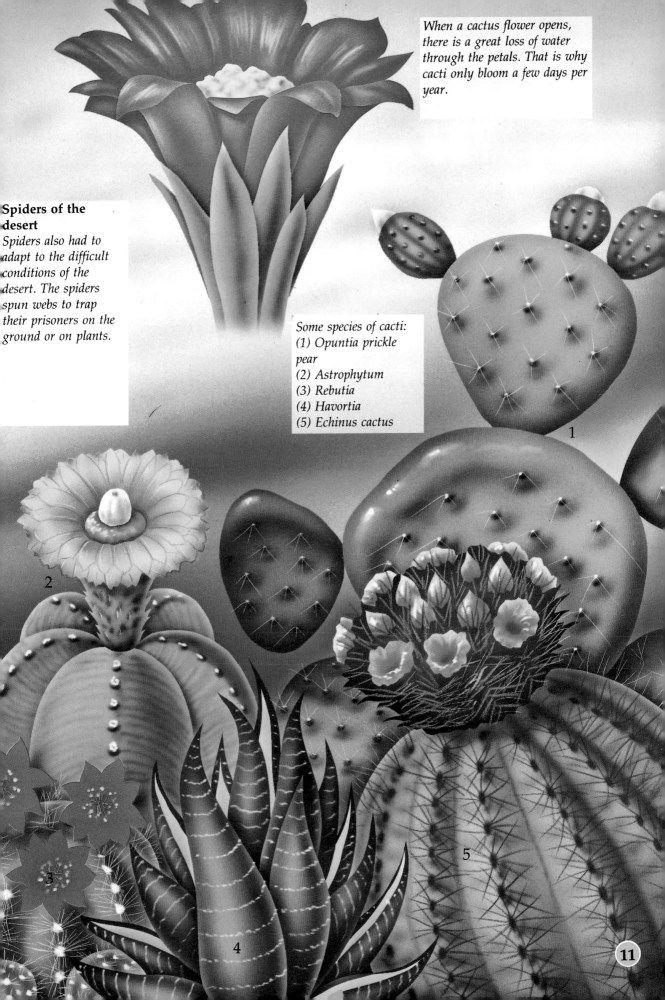

When a cactus flower opens, there is a great loss of water through the petals. That is why cacti only bloom a few days per year.

Spiders of the desert

Spiders also had to adapt to the difficult conditions of the desert. The spiders spun webs to trap their prisoners on the ground or on plants.

Some species of cacti:
(1) Opuntia prickle pear
(2) Astrophytum
(3) Rebutia
(4) Havortia
(5) Echinus cactus

THE FAMOUS SAGUARO CACTUS

The desert plant that you probably know best is the saguaro cactus, due more than anything to its characteristic shape and impressive size. It is over 49 feet (15 meters) tall, and its trunk can contain more than a ton of water.

Like all cacti, the saguaro grows very slowly. When it is 10 years old, it is still only 4 inches (10 centimeters) high, and at the age of 50 it only measures 6 feet (1.8 meters). The first branches, which grow vertically, do not appear until the plant is about 100 years old, which is when the cactus reaches maturity and begins to produce seeds.

At first, the seeds serve as food for rodents and other animals, and only one seed out of every 275,000 will ever become an adult saguaro. The saguaro keeps growing until it reaches a maximum height of about 66 feet (20 meters) and has about 50 branches. Normally, it reaches this height at the age of 250 years, which seems to be the maximum age of these cacti. An old saguaro can weigh up to 10 tons, of which about 1/5 is water.

Although very poisonous and covered by a multitude of thorns, the saguaro is a hospitable plant. Many different animals take advantage of the shade and cold surfaces that the saguaro provides, and some animals actually live inside the cacti themselves.

(1) A network of roots
The saguaro's roots are as long as the plant itself and extend over the surface in every direction. After it rains, the roots absorb the water and transport it to the elevated tissues.

(2) The saguaro's flowers
The saguaro produces beautiful white flowers which then become fruit, in the form of a sugary and nutritious pear— the cereus.

(3) Nests in the saguaro
When the saguaro reaches a good height, woodpeckers make a hole near the top and install their nests. The plant defends itself by surrounding the upper part of the trunk with sap, which hardens in order to avoid the loss of heat.

(4) Cactus owls
Years later, these same holes may be occupied by other birds, like the cactus owl, for example.

1

2

DESERT RAIN

It *does* rain in the desert, but it is never a normal rainfall; instead, when it rains, it is violent and torrential.

Desert rains are capricious: in some areas, not even one drop falls for several years (sometimes more than a decade) and yet, when it does rain, there is as much precipitation in a sudden deluge as would correspond to the rainfall for a whole year elsewhere. When desert rains arrive, water falls torrentially, and upon hitting dry land, rather than filtering down, it is rejected by the surface. Therefore, accumulated water often runs wild across the plains, is burned by the sun, or forms spectacular waterfalls in deep canyons. Water that has recently fallen carries a huge quantity of mud with it, so it barricades, breaks, and destroys everything in its path. Surprisingly, the precious liquid disappears almost as quickly as it falls, but its beneficial effects on the deserts are many.

Various seeds and animals remain in a latent state, perhaps for several years, asleep and hidden under the earth. So when the first important rain falls, the desert becomes a paradise populated by flowers and animals. Seeds are stimulated by a heavy downpour, and the water dissolves the strong external layer. Eggs and larvae hatch, insects buzz through the air, and frogs and other animals come out of their lethargy upon feeling the moisture.

In a few weeks, plants bud, bloom, and bear fruit. The development is very fast, but afterward the desert goes back to sleep until the next rain.

(1) Torrential rains
Desert rains can be very dangerous if the torrents overflow. However, even this event is useful to the plants, since many seeds are carried miles away safe in the beds of the torrents.

(2) Mud after the deluge
Finally, the water filters through the ground or reaches a dry lake. It seems impossible that not long after the deluge, all that is left is mud that dries out in the desert sun.

(3) Amphibians in the desert?
The North American horn toad uses its rear legs to bury itself under the desert, where it spends months waiting for rain. When the first puddles are formed, toads quickly come out mate, and lay their egg in the water. The small tadpoles grow very quickly, and become adults in just a few weeks, just in time to bury themselves before their puddle dries up completely.

(4) Desert flowers
After the rain, desert pea plants sprout and bloom for a few days.

3

14

THE DESERT BLOOMS

After the rain, a multitude of strong graminaceous stems, leguminous plants, and other small and medium-sized plants begin to appear on the sand as if by magic.

Most of these plants grow very quickly: the sprouts appear at night, and after a few days, the desert seems to be covered by a blanket of colored flowers that struggle to attract pollinating insects to their *corollas*, which are filled with nectar. They only have a few hours to complete the pollination process. Some species are capable, in an eight-hour period, of growing, blooming, fructifying, and then freeing their seeds and dying. These seeds may have to wait five years before they begin to grow.

New plants are not specially equipped to fight evaporation, and their roots are neither deep nor widely branched; this means that they are not at all well adapted for life in the desert. For these plants life is a race against the clock to bloom and produce seeds before the sun dries them out completely.

The seeds, however, are specialized since they contain the entire species' hopes for survival. They are covered by cuticles that need a certain amount of water to be dissolved, so they do not germinate until the second or third rainfall.

After a few weeks, the ephemerals have completed their life cycles and the desert once again appears to be empty. The new seeds will be transported by the wind until they become buried in the surface layers of the ground, and will stay "asleep" there, appearing to be dead, sometimes for more than three years.

(1) Millions of flowers
There are few sights that can compare to the vision of millions of flowers springing up from an empty desert after the first rainfall.

(2) So many seeds
Every year millions of seeds are produced that serve as nourishment to many animals in the desert, such as this small squirrel.

(3) A variety of shapes and colors
When they are completely grown, these plants are only made up of a flower and a short stem, which seems to spring directly up from the naked ground. Desert flowers can have quite varied shapes and colors, such as the giant carnegeia (3a) and the thick-leaved calandrina (3b).

(4) In the South American deserts
The puya is a succulent shrub typical of deserts in Chile and Peru.

(5) A North America flower
The shawii agave is a species of flower that lives in the deserts of California.

(6) Desert turtle
This turtle devours green leaves, fruit, and flowers, extracting half quart of water from the vegetable fiber, which it needs to survive the months of drought. The water is stored under it shell.

1

2

OASES

Did you know that deserts are not completely dry, although we may see very little water aboveground?

Almost all the water that exists in the desert is underground, and some of the watery layers are found at only 1 to 2 feet (30 to 60 centimeters) below the surface. But in some parts of the desert there are puddles and springs, and even great lakes holding millions of gallons of water.

Life forms take advantage of every possibility in the desert, so even around the smallest puddles there is always an oasis with lush vegetation. Here and there, surrounded by almost absolute dryness, green oases shimmer like attractive centers of life. Although it may surprise you, there are also rivers in the desert, but for most of the year they do not have any surface water.

However, often there is subterranean water circulating just below the apparently dry riverbed. Some vegetation grows year-round along many of these dry rivers, but when a storm comes and a flood follows, everything is dragged through the water, except those plants with the sturdiest roots.

If you ever travel through the desert and see trees, remember that they can only live relatively close to more or less permanent riverbeds. This means that in the surrounding area, even during the longest periods of drought, there are deep layers of water.

(1) Dry riverbeds
This is what the dry riverbed of a wadi looks like. As you can see, many plants profit by the water circulating underground. Oases spring up in the middle of the desert when the subterranean water level reaches the surface level.

(2) The date palm
It is one of the desert's treasures, since all of its parts are useful. Dates are a valuable food, the trunk's wood is used for construction, the base of the trunk is used as fuel, and the fibers that surround the trunk are used to make strong, resistent rope.

(3) Pin-tailed sand grouse
The adult pin-tailed sand grouse, carries water in its stomach and chest feathers. Thanks to this, the chicks (who feed on dry seeds) can survive many miles away from water.

(4) An artistic wasp
This single wasp mixes saliva with a little dirt to make nests in the form of a vessel. In each nest, the wasp places one egg, from which a great number of larvae hatch.

STUBBORN SHRUBS AND TREES

Are there really trees in the desert?

The answer is yes; one of the important life groups in the desert is formed by atrophied trees and bushes. They are normally found in sandy deserts, although they are present in almost all desert areas.

Trees and shrubs in the desert get water from deep underground, thanks to their very long roots, which can reach up to 246 feet (75 meters). Also, these roots keep the plants held fast to the surface, especially if they manage to get hold of a solid rock.

To limit evaporation, many of these shrubs and trees have very few leaves if any at all. When the driest season arrives, they lose their leaves and some of their branches and breathe through the porous bark of their trunks. The woody tissue of the trunk and main branches are very strong and resistent to the effects of the sun, and some bushes can dry out all the way without dying.

So, when the first rains begin to fall, leaves and new sprouts begin to grow out of the branches that seemed to be dead. As you might guess, the plant's functions are greatly reduced when it loses its leaves, and they can stop altogether. During the hottest part of the year, therefore, the plant maintains a lethargic state. The result of this lifestyle is a very slow growth and a very long life.

The barrel cactus
According to legend, the barrel cactus contained enough water to save the life of a lost traveler. Its shape changed whether it was full or empty.

The saxaul
The saxaul looks li[ke] a 13- to 16-foot (4-5-meter) bush. As you can see, in the summer the white saxaul looks like a dead tree, since it loses its leaves and almost all of its branches.

The desert mallee

These are mallee desert flowers. Desert flowers are some of the most beautiful in the world, since they have to attract insects in very little time.

Nourishment of excrement

This desert beetle makes a ball with excrement from camels and goats and rolls it to its underground nest, where its babies are. There they can feed on the ball for several weeks.

Deformed by the wind

The wind's force deforms some desert trees and bushes so they adapt themselves in order to have the smallest surface possible facing the wind. Have you noticed what strange shapes they sometimes have?

21

THE DREADFUL LOCUST PLAGUE

Around the Sahara Desert, there exists an unforeseeable enemy who, in a few hours, can destroy all the vegetation and ruin entire crops: it is the migratory locust, one of the most dreadful plant-destroying insects around.

Locusts live and reproduce at the edge of the desert, and in some species the eggs can stay in a lethargic state underground for up to three years, waiting for the first rains to go through eclosion, although most of them hatch after three or four weeks.

However, when the individual density reaches a certain point, there comes a big change. Locusts change shape, color, physiology, and overall behavior. That is how the lone, harmless locust becomes, in one or two generations, a dreadful devastator. Young locusts without wings form very dense groups that move around in tight packs (up to 20,000), walking on the ground; the adults, on the other hand, are equipped with wings and take flight together in dense clouds that may even block the sun's light. Keep in mind that a swarm can be made up of millions and millions of individuals, up to 40 billion locusts! When the cloud descends, it devours all the vegetation within a huge expanse of up to 400 square miles.

Locust clouds have extraordinary mobility, and some of them can travel thousands of miles in search of food. The longest trips are up to 3,000 miles.

(1) A deadly cloud
When the cloud lands, it causes tremendous damage to the vegetation, given that it can devour more than 40,000 tons of vegetation per day. Neither young locusts nor adults need to drink, since they get water from the plants they eat.

The life cycle of locusts
2a) Up to 1,000 locusts can lay eggs in the same square mile of land.
2b) Locust larvae.
2c) Young hoppers: their legs are so powerful that they can jump up to 10 times their own length, so they can escape from predators, like serpents.
2d) Adult locusts: the more they grow, the hungrier they are, and once they grow wings, they devour all the vegetation within their reach.

2d

2c

2b

PLANTS THAT ABSORB MIST

Some plants and animals of the desert take advantage of the scarce existing water, because they are able to drink the dew.

In many coastal deserts, the ocean's mist provides humidity that is indispensable for existence. In other deserts, including some very far from the sea, when humidity hits a cold surface, a type of dew that covers rocky surfaces in the mornings is formed.

In the Namib Desert on the west coast of South Africa, one of the rarest plants in the world exists— the welwitschia mirabilis—which is practically a living fossil.

It has a very strange shape, since it is really a crushed *conifer* and has a very thick root, which can measure 3 feet wide. The upper part of the root sticks through the ground's surface, and the leaves of the plant come out of it. The leaves can grow to be hundreds of feet long, but the difficult conditions of life in the desert continuously break the leaves, and they rarely exceed a few feet in length.

The welwitschia needs so much surface space because it gets water by capturing it from the fog, by way of its leaves. Sea fog produces drops that condense on the surface of the leaves, and they absorb the moisture. Then, the water is transported through a network of tubes toward the roots, where it is stored.

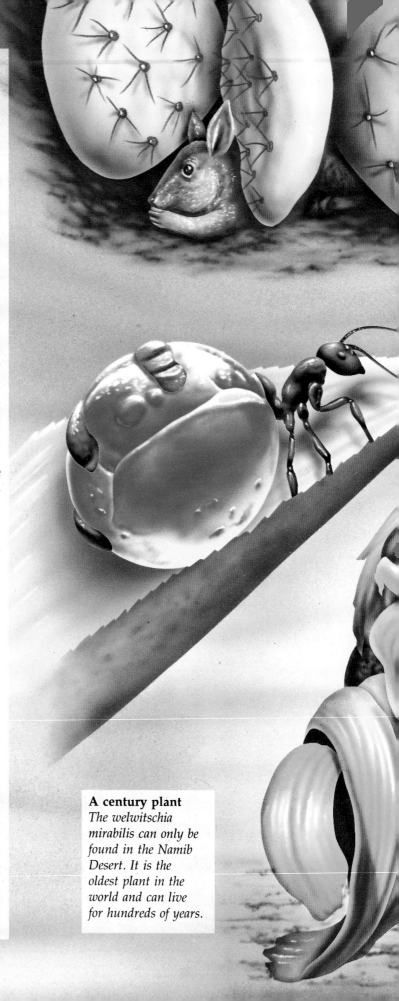

A century plant
The welwitschia mirabilis can only be found in the Namib Desert. It is the oldest plant in the world and can live for hundreds of years.

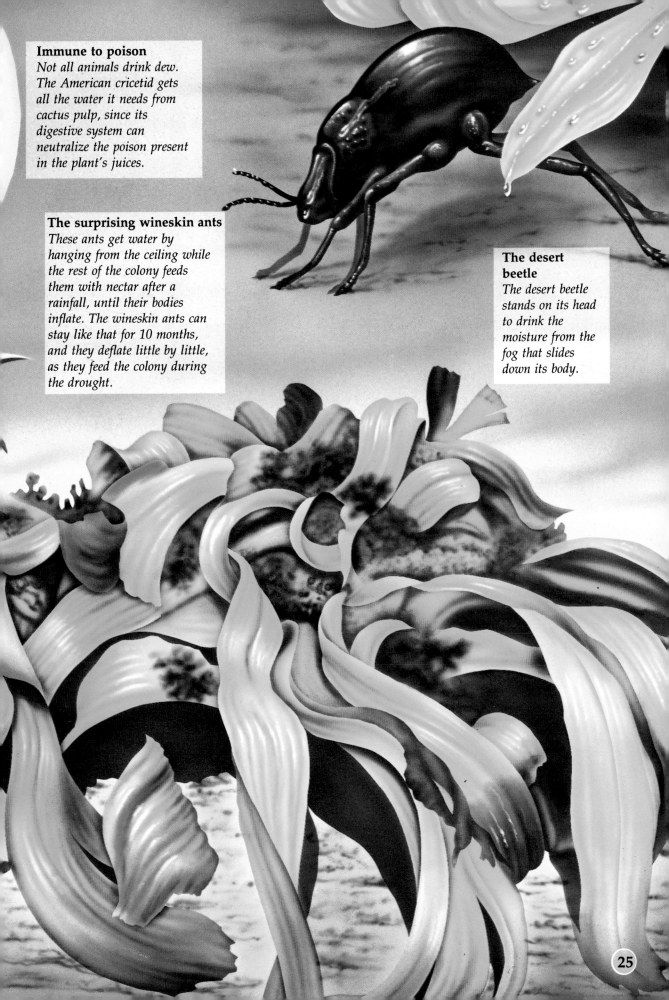

Immune to poison
Not all animals drink dew. The American cricetid gets all the water it needs from cactus pulp, since its digestive system can neutralize the poison present in the plant's juices.

The surprising wineskin ants
These ants get water by hanging from the ceiling while the rest of the colony feeds them with nectar after a rainfall, until their bodies inflate. The wineskin ants can stay like that for 10 months, and they deflate little by little, as they feed the colony during the drought.

The desert beetle
The desert beetle stands on its head to drink the moisture from the fog that slides down its body.

CONFRONTING THE HEAT OF THE DESERT

One of the biggest worries for plants and animals in the desert is maintaining adequate body temperature. It is so hot!

Desert plants use mechanisms that avoid the loss of water and can survive losing a greater amount of water than plants in other environments. The tamarugo carob tree, for example, is a tree that lives in the Atacama Desert in Chile. It is able to withstand 10-year droughts between one flood and another.

Animals, on the other hand, have a big disadvantage compared to plants: most animals need water frequently. To combat this, some desert animals like the camel can store water in their bodies and therefore survive long periods without drinking.

Other animals get all the water they need from vegetation, including even the driest seeds. Predators obtain most of the water they need from the blood and other fluids of their victims.

However, animals do have one very important advantage over plants: they can move to hide themselves from direct sunlight. Because of this, they can decide which part of the day they want to spend active, and where they are going to spend the hottest hours.

(1) Protected from the sun
An enormous expanse of sand in the burning sun leaves animals and plants to "think" of ways to survive in the inhospitable desert.

(2) A safe hole
The neotome sleeps safely in his den, which is a superficial hole covered with pulpy and dead leaves and pieces of dry branches.

(3) Protective shade
Shrubs provide shade for all sorts of animals, from hares to coyotes and antelope.

(4) Practical defer
The butcherbird use thorns to hang the prisoners it capture like this poor desert mouse.

(5) Protecting itse with thorns
In the first stages o growth, the cactus totally covered in thorns, since this is when it runs the m risk of being devou

(6) A good refug
Many beetles and o insects stay cool an the stems of plants, since there the temperature is only degrees Fahrenheit.

THE CREOSOTE: A MILLENNIAL SHRUB

The creosote shrub, named for its characteristic odor, grows in American deserts.

The creosote is actually one of the most common desert plants. It survives the difficult desert conditions thanks to its impressive system of roots, which is so extensive that it can absorb even the last drop of moisture on the surface layer of the ground. Consequently, no other plant lives near it, and creosote bushes are normally surrounded by a circle of naked terrain.

Some of these bushes have developed such an effective root network that the plant has kept growing for thousands of years. As it grows, the shrub forms an empty circle at the rate that the oldest branches in the center die. The biggest bushes can get to be more that 82 feet (25 meters) in diameter, and although their most visible branches are "only" a few hundred years old, the bush can be more than 1,000 years old.

In order to secure all the water in the area, the creosote's roots secrete actual poison that kills the seeds of neighboring plants and impedes the possibility of invading roots penetrating its vital space.

A bad neighbor
The creosote does not let any other plant grow near it. There is a large separation between the plants.

Fight to the death
1) Wasps feed on flower nectar, but they must find a tarantula to reproduce. 2) When the two rivals meet, a fight to the death ensues. 3) The wasp buries the tarantula after having deposited an egg on its abdomen. When the larvae are born, they slowly devour the tarantula, which is still alive.

Gila monster
Some very strange animals live in the desert, such as this lizard called Gila monster. It is one of the only poisonous lizards that exists.

DAY AND NIGHT IN THE DESERT

At night the desert's temperature drops rapidly, since the absence of clouds means that nothing retains the ground heat.

Even in the hottest deserts, nights are very cold and temperatures can drop below 32 degrees Fahrenheit. The drop in temperature has one advantage: it lowers enough so that some of the water vapor in the air is left behind, because of the *condensation*, in the form of dew. As you have seen before, this tiny amount of water is vital for some forms of desert life. On the other hand, some plants have adapted specially to nighttime conditions, like cacti, which only open their pores at night, when the loss of water through transpiration is smallest. There are even some species that only open their flowers at night. And among animals, one strategy for avoiding the heat of the day is to only be active at night.

For this reason, many desert animals are *nocturnal* and spend the daytime hidden in the shade or underground. The animals that are active during the day have developed methods to avoid an uncontrolled increase in body heat. Some of them, like scorpions or desert squirrels, alternate active periods in the sun with periods in the shade when they cool down.

(1) When the sun hides
When the sun sets on the sand of the desert, the animals and plants can employ other means to survive.

(2) Night flower
The cereus flower can only be observed in the darkness because it completes its entire cycle in just one night.

(3) Scarce food
The jerboa eats the seeds, roots, and plants that it can get. When it eats, sometimes it sits on its tail and back legs, like a kangaroo.

(4) Water deposits
The neotome gnaws at the nopal prickly pear, which is 80 percent water, to get the water it needs to survive.

(5) Food and water
The bulbs and tubers of perennial grasses are constantly searched for by rodents and other animals, like these peccaries that dig in the sand to find in their pulp not only food, but also the stored liquid.

4

GLOSSARY

cactus a plant found in the desert that has fleshy stems and branches with thorns instead of leaves

condensation a chemical reaction in which steam changes into water

conifer a family of trees including evergreens and shrubs, some of which have pinecones or fruit

corolla the second inner circle of a flower, made up of transformed, generally colored, leaves called petals

dunes hills or ridges of sand that are piled up by the wind

nocturnal a plant or animal that is only active at night

photosynthesis a process in which green plants synthesize organic material through carbon dioxide, using sunlight as energy

phytophagous insects insects that feed off plants

shoots stems or branches with leaves and appendages, usually not yet mature

INDEX